Bacteria and
ARCHAEA

Rebecca Woodbury, Ph.D., M.Ed.

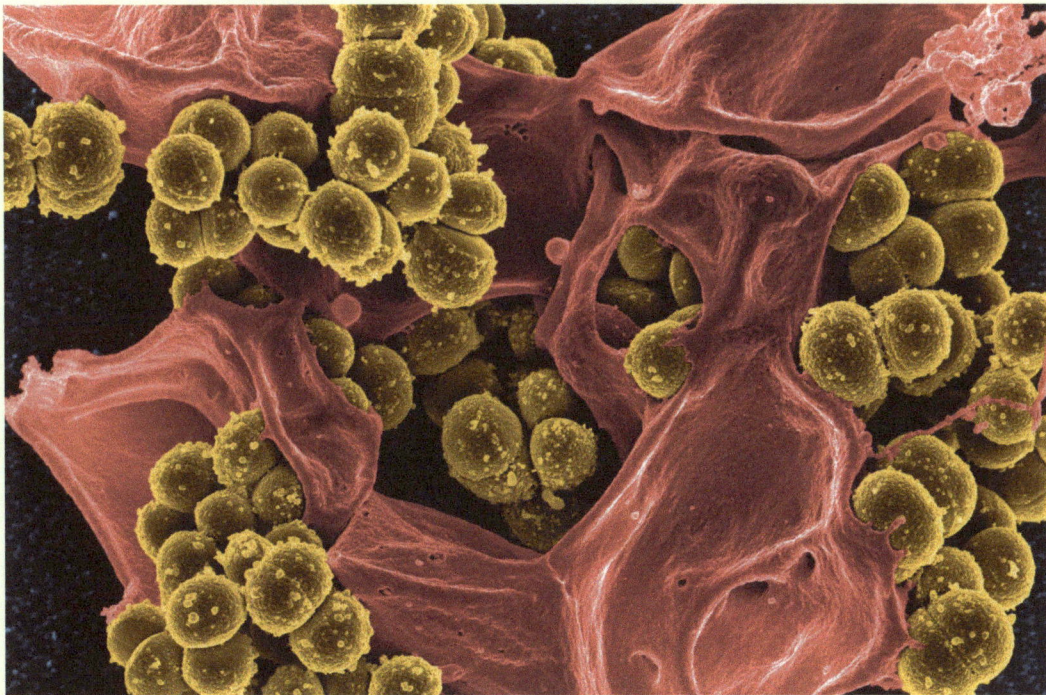

Gravitas Publications Inc.

Bacteria and
ARCHAEA

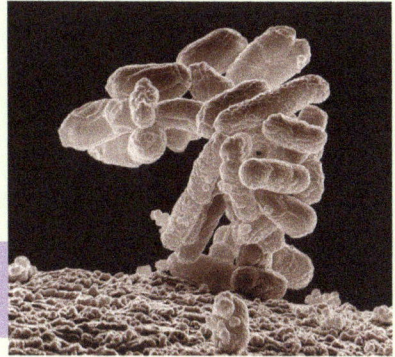

Illustrations: Janet Moneymaker

Bacteria and Archaea
ISBN 978-1-950415-51-9

Published by Gravitas Publications Inc.
Imprint: Real Science-4-Kids
www.gravitaspublications.com
www.realscience4kids.com

RS4K

Credits: Cover, Title Pg, P.9. National Institute of Allergy and Infectious Diseases (NIAID)/NIH; Above: Public Domain; P.3. Public Domain, NPS; P. 7. Inset: Darryl Leja, National Human Genome Research Institute, National Institutes of Health; P.15 Top, Алексей Синельников-AdobeStock; Bottom, Public Domain; P.17. - 1. CDC/Rob Weyant; 2-3. Public Domain; P.21. NOAA

Have you wondered what
might live in really hot places?

Something lives in there?

Grand Prismatic Spring, Yellowstone National Park, USA

Have you wondered
what might live in really
smelly places?

I would not
live there!

Did you know your gut
is home to millions of
tiny **organisms**?

Does my stomach look like that?

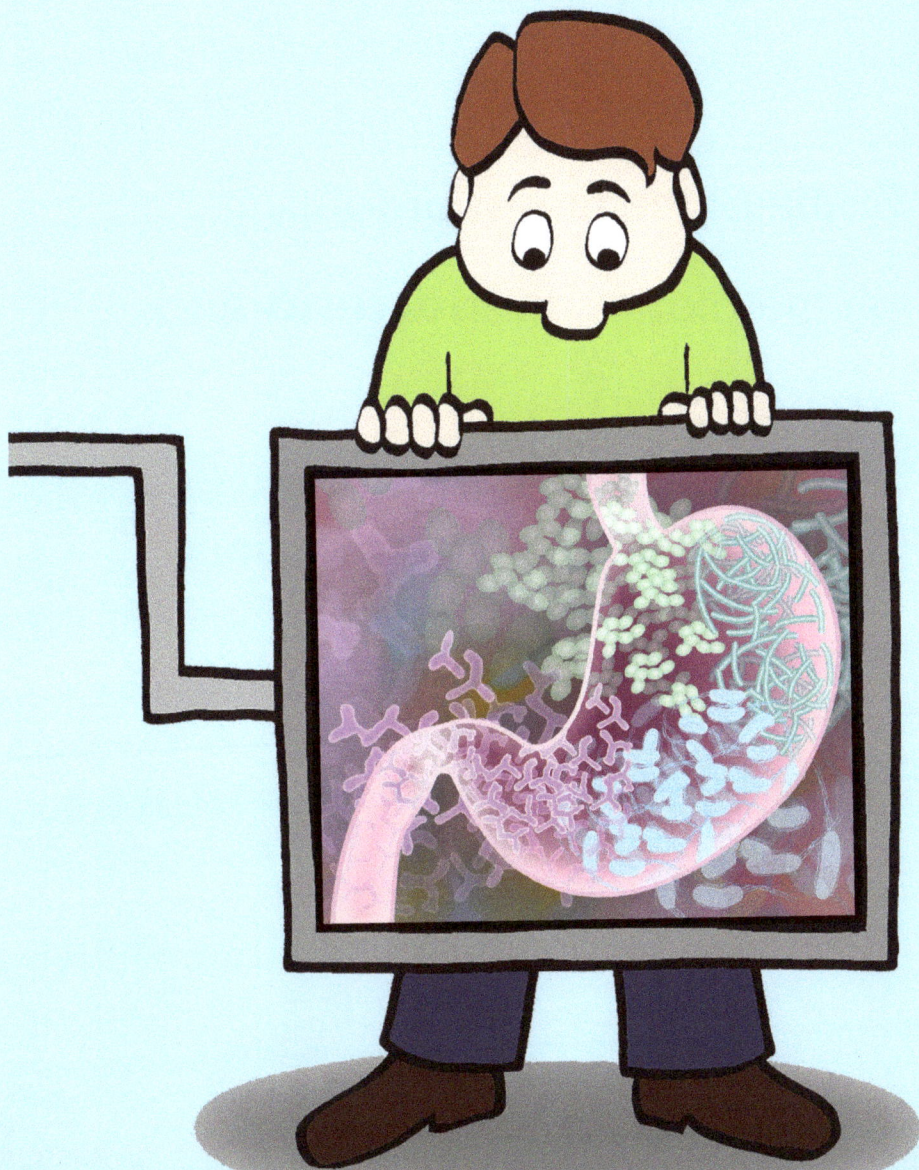

Some of the tiniest organisms on the planet are **bacteria** and **archaea.**

They are so tiny that you cannot see them with only your eyes.

You need a microscope to see this!

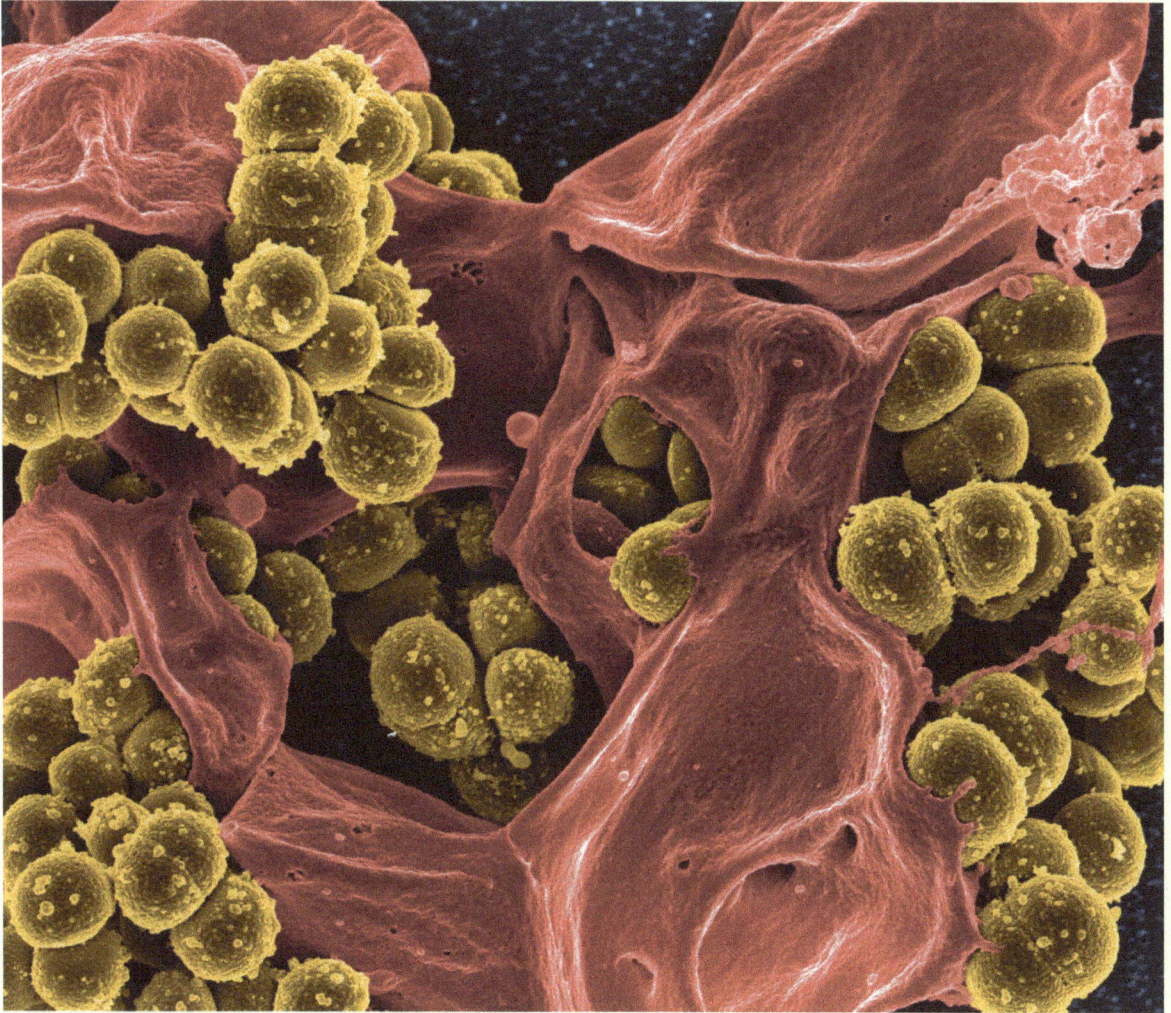

The yellow spheres are bacteria.

Bacteria and archaea look a lot alike.

In the past, scientists put bacteria and archaea in the same group.

Today, scientists know that they are different from each other.

They look like they could be the same.

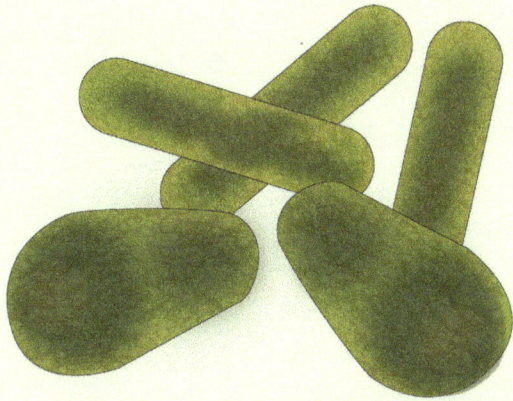

Archaea

Bacteria

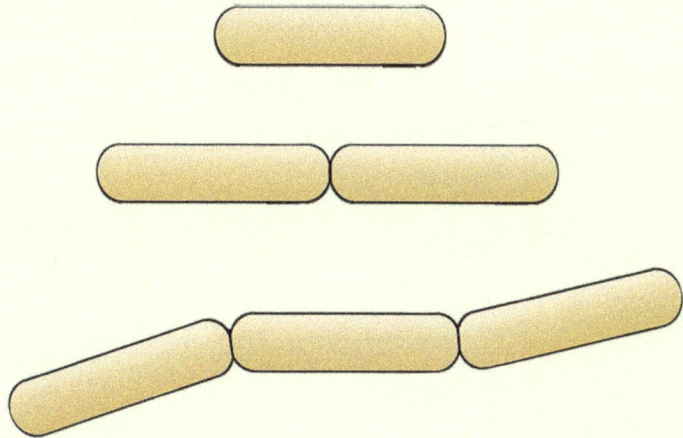

Bacteria have three basic shapes:
rods, spheres, and **spirals.**

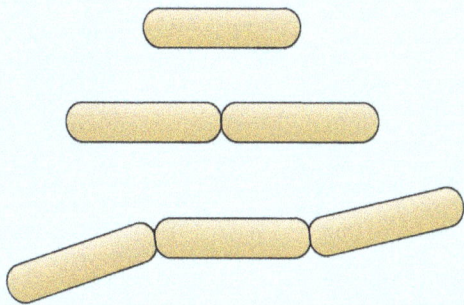

Rod-shaped bacteria look like short hot dogs.

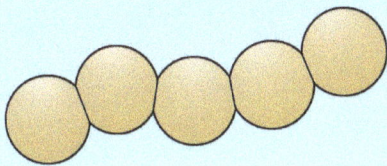

Sphere-shaped bacteria look like tiny ping pong balls.

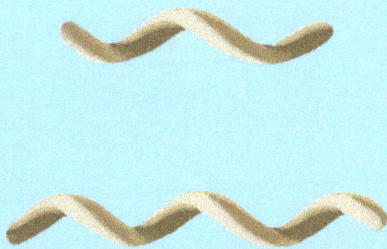

Spiral-shaped bacteria look like wiggly snakes.

Bacteria can live in many different places.

Some bacteria live in dirt.

Some bacteria live in water.

Can you live in different places?

Some bacteria live alone.
And some bacteria live in a
big group called a **colony.**

Colonies of bacteria can be big enough to see with only our eyes!

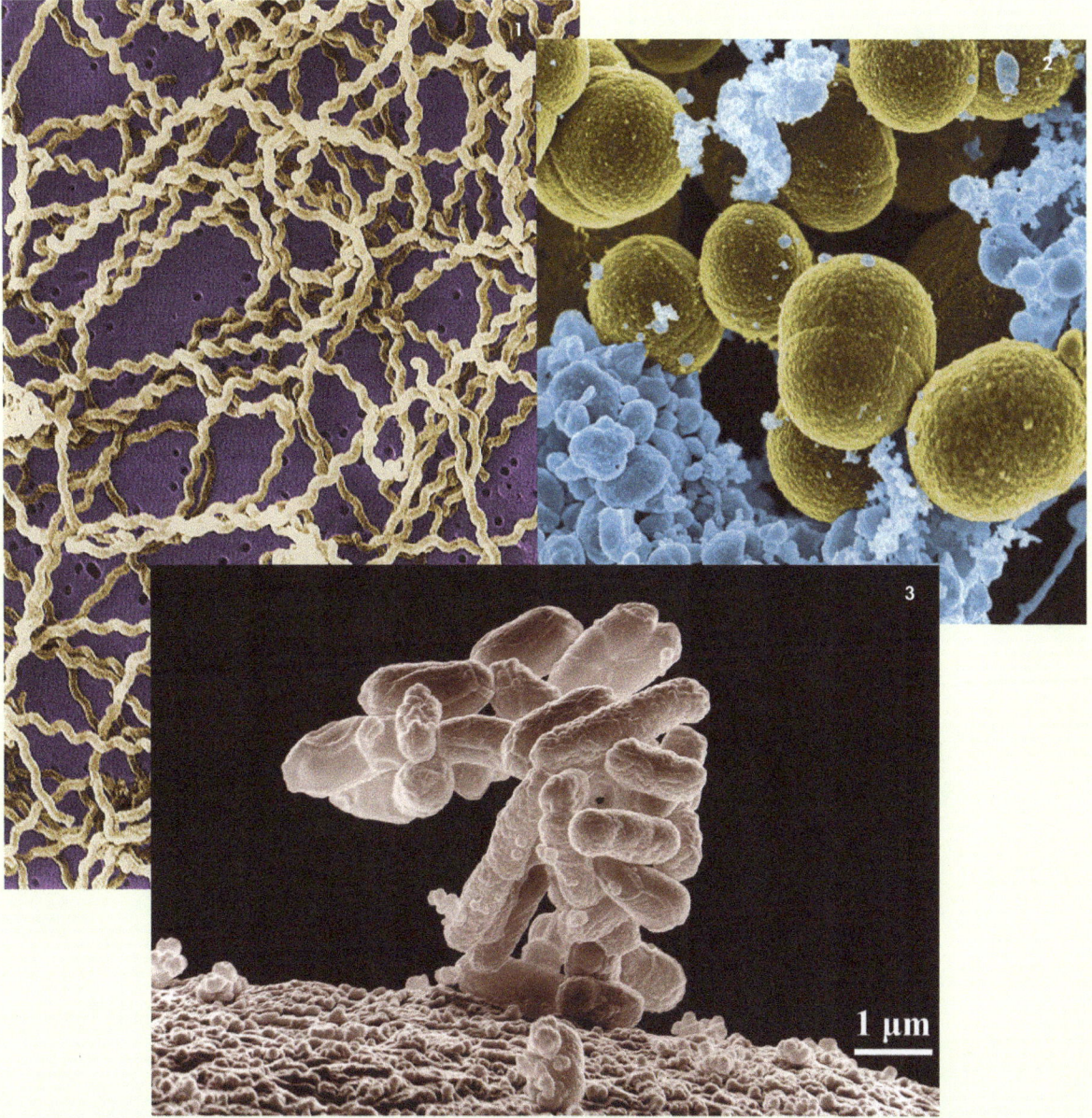

1 μm

Like bacteria, archaea
also have different shapes.

Some archaea look like
a ball with lots of squiggly
hairs on one side.

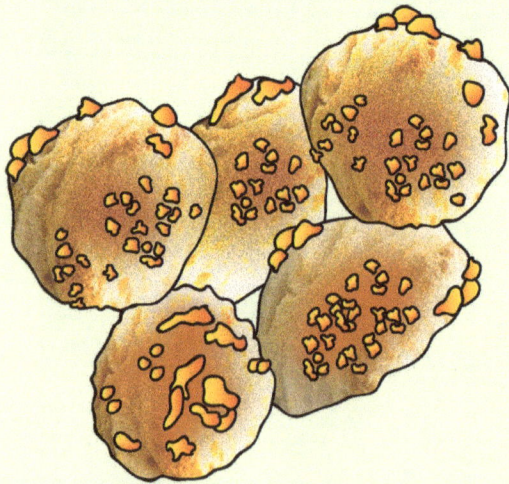

Some archaea
look like lumpy
ping pong balls.

Some archaea are rod shaped.

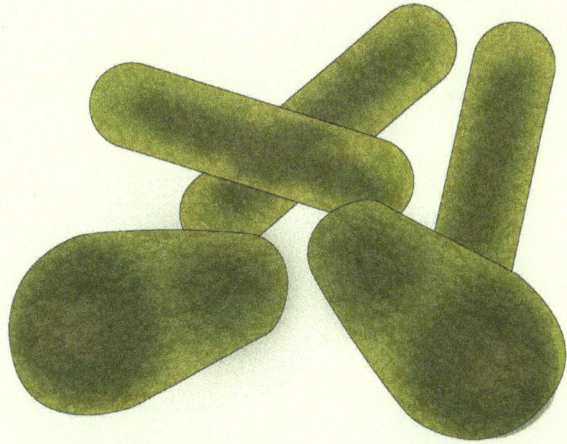

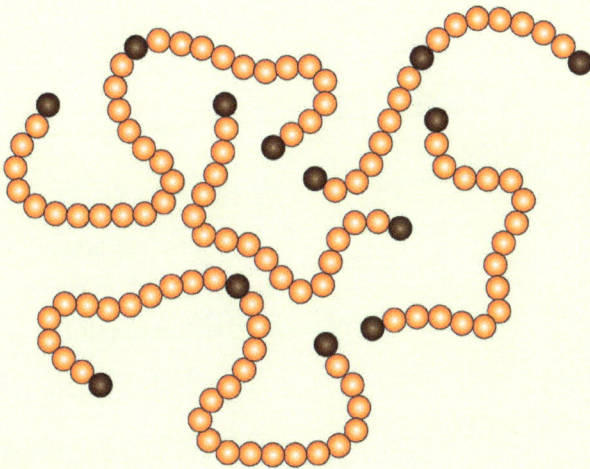

Some archaea form long hair-like shapes.

Archaea can be found in very
hot and very salty places.
Some archaea even eat iron!

Archaea live in
steaming volcano
vents in the ocean!

How to say science words

archaea   (ahr-KEE-uh)

bacteria   (baak-TIHR-ee-uh)

colony   (KAH-luh-nee)

organism   (AWR-guh-nih-zuhm)

rod   (RAHD)

sphere   (SFEER)

spiral   (SPIY-ruhl)

www.ingramcontent.com/pod-product-compliance
Lightning Source LLC
Chambersburg PA
CBHW040148200326
41520CB00028B/7536